IDEA SINGULARITIES

INTRO----------------------p. 5

Singularity Inventions-------------------p. 7

Matchik Inventions----------------p. 29

Tractatus of Strategic Knowledge ------p. 39

Singularity Spells (Occult Calendar) --------p. 77

Notes on Immortality, Time-Travel, Teleportation, Telekinesis
---p. 95

BIO --------------------------p. 110

Dedicated to those nifty problems

one needs an excuse to solve

© 2004, 2013, 2014, 2015, 2016, 2017, 2018, 2019, 2020, 2021 Nathan Coppedge. Some material previously available for free under citation of the same author.

IDEA SINGULARITIES: BLACK SWAN INVENTIONS IN AN AGE OF HIGH TECHNOLOGY

BY NATHAN COPPEDGE

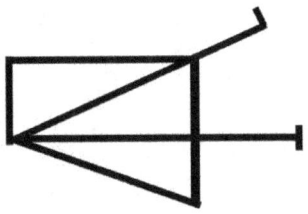

(…)

Intro

This text deserves no long introduction. It is simply a collection of singularity-related novelties and inventions.

Quite of few of these ideas could come in practical use.

The aim is to inspire a prolonged golden age of human discovery.

Although most of this content is available on the internet, Coppedge still controls the copyright and proper citation is required in any re-use.

You may be a bit terrified with the variety of work which I have acquired over the years almost all by myself.

It has been quite an interesting experience and I hope my readers continue to enjoy my work for years to come!

Maybe mere novelty is not what I mean.

(…)

§1: SINGULARITY INVENTIONS

I would advise that you may resist it, but you cannot fully overcome it.

CONVENTIONAL WAVEFORMS REINCLUDED CORE

X UNIVERSE 0 D = 11 IMPOSSIBILITY WAVE RESULTS= INF, EFF= NEG INF, DIFF= IMPOSSIBLE	X UNIVERSE 5 D = 6 RARITY WAVE RESULTS= FIN, EFF= NEG INF, DIFF = INF	UNIVERSE 10 D = 4 IDEA WAVE RESULTS= 0, EFF= NEG INF, Diff = INF	X UNIVERSE 15 CLEAR WAVE D = 2 RESULTS= NEG FIN, EFF= NEG INF. DIFF = INF	UNIVERSE 20 D = NEG1 INSANITY WAVE RESULTS= NEG INF, EFF= RESULTS DIFF= 0
UNIVERSE 1 D = 10 COHERENT WAVE RESULTS= INF, EFF= NEG FIN DIFF= RESULTS	UNIVERSE 6: SOULS D = 0 RESULTS = FIN EFF = NEG FIN DIFF = -(EFF) + RESULTS	UNIVERSE 11: SKILLS RESULTS = ZERO EFF = NEG FIN DIFF = FIN	UNIVERSE 16: TIME D = 0 RESULTS = NEG FIN EFF = NEG FIN DIFF = -(EFF)+RESULTS	UNIVERSE 21: D = NEG2 FRINGE WAVE RESULTS= NEG INF, EFF= NEG FIN. DIFF = RESULTS
UNIVERSE 2 D = 9 GROWTH WAVE RESULTS=INF, EFF= 0, DIFF= RESULTS	UNIVERSE 7: EMOTION D = 0 RESULTS = FIN EFF = 0 DIFF = RESULTS	UNIVERSE 12: DRUGSETC D = 0 RESULTS = ZERO EFF = ZERO DIFF = ZERO	UNIVERSE 17: DAMAGE D = 0 RESULTS = NEG FIN EFF = 0 DIFF = RESULTS	UNIVERSE 22: D=NEG3 CATEGORY WAVE RESULTS= NEG INF, ····, DIFF = RESULTS
UNIVERSE 3 D = 8 STANDING WAVE RESULTS= INF EFF = FIN, DIFF= RESULTS	UNIVERSE 8: CORE D = 0 RESULTS = FIN EFF = FIN DIFF = RESULTS - EFF	UNIVERSE HUMANS RETRACTORS D = 0 RESULTS = AVG ZERO EFF = FIN DIFF = AVG 0 - EFF	UNIVERSE 18: EFFECTS RESULTS= NEG FIN EFF = FIN D = 0 DIFF= NEGEFF+RESULTS	UNIVERSE 23: D = NEG4 LUX WAVE RESULT = NEG INF, EFF= FIN, DIFF= RESULTS ATEGORY
UNIVERSE 4 D=7 DISINTEGRAL WAVE RESULTS= INF, EFF= RESULTS DIFF= 0	X UNIVERSE 9 D = 5 FUNCTION WAVE RESULTS= FIN, EFF= INF, DIFF= NEG INF	UNIVERSE 14 LUCK WAVE RESULTS 0, EFF= INF, D = 3 DIFF= NEG INF	UNIVERSE D = 1 19 COMMUNIC WAVE RESULTS= NEG FIN, X EFF= INF, DIFF= NEG INF	X UNIVERSE 24 D = NEG5 OPPORTUNITY WAVE RESULTS= NEG INF, EFF= INF, DIFF= IMPOSSIBLE

Free and must remain non-proprietary, Nathan Larkin Coppedge

POSSIBILITY WAVES

Universe 1: Coherent Wave, Wishing [10 - Dimensional]

Universe 2: Growth Wave, TOE [9 -Dimensional]

Universe 3: Standing Wave, Perpetual Motion [8 - Dimensional]

Universe 4: Disintegral Wave, Mathematical Integration [7 -Dimensional]

Universe 5: Rarity Wave, Paradoxes [6-Dimensional]

Universe 6: Souls, Psychic Source [0-Dimensional]

Universe 7: Emotions, Product of Damages [0-Dimensional]

Universe 8: Cores, Major Technology [0-Dimensional]

Universe 9: Function Wave, Equal Opposites [5-Dimensional]

Universe 10: Idea Wave, Energy Produced [4-Dimensional]

Universe 11: Skills [0-Dimensional]

Universe 12: Zeroverse, Drugs [0-Dimensional]

(POSSIBILITY WAVES, CONT'D)

Universe 13: Retractors, Humans [0-Dimensional]

Universe 14: Luck Wave, Immortals [3-Dimensional]

Universe 15: Clear Wave, Languages [2-Dimensional]

Universe 16: Timed Events [0-Dimensional]

Universe 17: Damages, Warping [0-Dimensional]

Universe 18: Effects, Major Results [0-Dimensional]

Universe 19: Communication Wave , Education [1-Dimensional]

IMPOSSIBILITY WAVES

Universe 0: Impossibility Wave, also called Emotion Wave, Psychic Wave [11-Dimensional]

Universe 20: Insanity Wave also called Perfection Wave, Soul Wave [-1-Dimensional]

Universe 21: Fringe Wave also called Standardization Wave, Exception Wave [-2-Dimensional]

Universe 22: Category Wave also called Aesthetic Wave, Beauty Wave [-3-Dimensional]

Universe 23: Luxury Wave, also called Reality Wave, Time Wave [-4-Dimensional]

Universe 24: Opportunity Wave, also called Creativity Wave, Working Wave [-5-Dimensional]

Impossibilities were collapsed Nov 2021 due to a more accurate calculation of trans-finites.

[Added Nov 24, 2021. See: Universe Profile]

...

Mechanics can now be improved virtually at will:

PMMS: RESULTS FOR THE SPECIAL VALUE THEOREM

DIFF	EFFICIENT	HIGH EFF	ADVANCED
5	2.5	3.5	4.5
4	2	3	4
3	1.5	2.5	3.5
2	1	2	3
1	0.5	1.5	2.5
0	0	1	2
-1	-0.5	0.5	1.5
-2	-1	0	1
-3	-1.5	-0.5	0.5
-4	-2	-1	0
-5	-2.5	-1.5	-0.5
	EFF 2	EFF 3	EFF 4	EFF

```
 2    ---------------- FLYING
 1.5  -----WEIGHT AND BALANCE
 1    ---------------- BALANCE
 0.5  ---------------- WEIGHT
 0    ---------------- NOTHING
-0.5  ---------------- BUOY
-1    ---------------- BUOY BALANCE
-1.5  -----BUOY BALANCE AND BUOY
-2    -----EXPONENTIAL GRAVITY LEVER
```

There is a formula for the contents of any book based on the title:

SOUL OF LITERATURE FORMULA

Title of book = '[quality of X] [opp qualifier]'

Soul of the book = 'If you [X] qualifier [subject of X and qualifier] [opp X clarified]'

—How do I find the soul of literature?

The Top 20 works of every major intellectual can be looked up with relative ease:

GREAT PHILOSOPHY HISTORICAL MODEL BY NATHAN COPPEDGE

What is obvious? [input]

Opposite of obvious? [input]

What is trivial in this time? [input]

Pathetic argument that might win? [input]

What is the better 2-step of [trivial]?

WISE ANSWER? [input]

What is most required for [trivial]???

You will find it is [WISE ANSWER]

PRIMARY INVENTION [WISE ANSWER]

That wishes for [trivial]

Philosopher is remembered as studying [Opposite of obvious]

MAJOR WORKS OF PHILOSOPHY

MAJOR WORK 1: [Opposite of obvious] application of [WISE ANSWER].

MAJOR WORK 2: Theory missing [trivial]

MAJOR WORK 3: In more than one way [trivial] is [obvious]

MAJOR WORK 4: [trivial] is also [opposite of obvious]

MAJOR WORK 5: [obvious] IT IS... BUT IT IS ALSO [opposite of obvious]

MAJOR WORK 6: Variations on concepts of [trivial]

MAJOR WORK 7: Theories about theory missing [trivial]

MAJOR WORK 8: [Opp of obvious] is missing something!

MAJOR WORK 9: Not [Obvious] with [Wise answer]

MAJOR WORK 10: [Wise answer] is great

MAJOR WORK 11: Wishing for [Trivial] is not [Obvious]

MAJOR WORK 12: What is not [Obvious] is [Wise answer]

MAJOR WORK 13: [Trivial] is missing, a theory missing [Trivial]

MAJOR WORK 14: A theory of [Trivial] is not a theory

MAJOR WORK 15: [Trivial] beyond [Trivial] beyond [Trivial]

MAJOR WORK 16: Beyond [Trivial] IS [Opp of Obvious]

MAJOR WORK 17: Paradoxical [Opp of Obvious]

MAJOR WORK 18: [Trivial] IS paradoxical

MAJOR WORK 19: Paradoxical [Obvious]

MAJOR WORK 20: [Wise answer] transcends reality

Higher Art Form: [opposite of obvious] WITH [trivial]

Recently, one may research known and unknown languages MATHEMATICALLY by the number of DIMENSIONS OF THE LANGUAGE:

2-d: 1.585

One and a half.

And 8 / 100.

And 1 / 200.

3-d: 1.72765

One. 7/10 About 3/100ths

Minus 3/1000s

And about 2/3 10,000s

Minus 1 / 100,000

4-d: 2.9731385

Three. Minus 2 / 100.

And minus 6.8615 / 1000.

Even more recently, one can predict future ideas by following a regular cyclical pattern and synchronizing one or more recent inventions with the pattern.

HISTORY OF IDEAS PAPER
START ANYWHERE, ARRANGE CHRONOLOGICALLY
These refer to rough dates of each invention as a science.

Technological Complex is	Technological Complex is
Technological Simple is	Technological Simple is
Artistic Simple is	Artistic Simple is
Artistic Complex is	Artistic Complex is
Cosmological Complex is	Cosmological Complex is
Cosmological Simple is	Cosmological Simple is
Physical Simple is	Physical Simple is
Physical Complex is	Physical Complex is
A New Concept is	A New Concept is
Technological Complex is	Technological Complex is
Technological Simple is	Technological Simple is
Artistic Simple is	Artistic Simple is
Artistic Complex is	Artistic Complex is
Cosmological Complex is	Cosmological Complex is
Cosmological Simple is	Cosmological Simple is
Physical Simple is	Physical Simple is
Physical Complex is	Physical Complex is
A New Concept is	A New Concept is

CHARACTERISTICA UNIVERSALIS

MAIN CHARACTERISTICA:

- Categories.
- Vertical = entity, value, principle, power.
- Horizontal = degree, standard, commonality, honor.
- Diagonal: judgment, energy, resources, substance.
- Organic lines: coherence, boundary, dimensions, limit.
- Systems: identities.
- Substance: quanta, bosons, spacetime, posits.
- Abstracta: complexity, efficiency, perfection, beauty
- Organon: Nature, Wisdom.
- Flags: Inflection, Incorporation, Notation, Tradition.

—Characteristica Universalis (...)

ENERGY EQUATION FOR ORDINARY OBJECTS VERSUS PERPETUAL MOTION

FOR ORDINARY OBJECTS

[(MIN EFF + 1) - (MAX EFF + 1)] / [0.5 (MIN EFF + MAX EFF)]

FOR PERPETUAL MOTION:

[(MIN EFF + 1) - ((MAX EFF / 2) + 1)] / [0.5 (MIN EFF + MAX EFF)]

This is an accurate formula for perpetual motion (/2 simply means multiplying by 0.5 for a level gradient, a slight slope ends up being * 0.51 to * 0.68 or more which reduces the overall rating slightly)

IMPROVED COMPUTER ARCHITECTURE

GIVEN QUESTION (A) is C: B-D SOUL is BCAD or DCAB

GIVEN QUESTION (B) is D: C-A SOUL is CDBA or ADBC

GIVEN QUESTION (C) is A: D-B SOUL is DACB or BACD

GIVEN QUESTION (D) is B: A-C SOUL is ABDC or CBDA

Arrange so category opposite of A is in position C, and category most opposite of B is in position D. Opposites may be used instead of given data.

—Universe DNA (...)

TOE TECH PRINCIPLES

- Math + TOE.
- Wish + Perpetual Motion.
- TOE + Elements.
- Perpetual motion + Meaning.
- Elements + Function.
- Meaning + Energy.
- Function + Variation.
- Energy + Language.
- Variation + Psychic.
- Language + Organization.
- Psychic + Species.
- Organization + Set.
- Species + Resources.
- Set + Sufficiency.
- Resources + Math.
- Sufficiency + Wish.

TOE Tech Principles (...)

...

SKETCH OF TECH UP TO 15TH DIMENSION

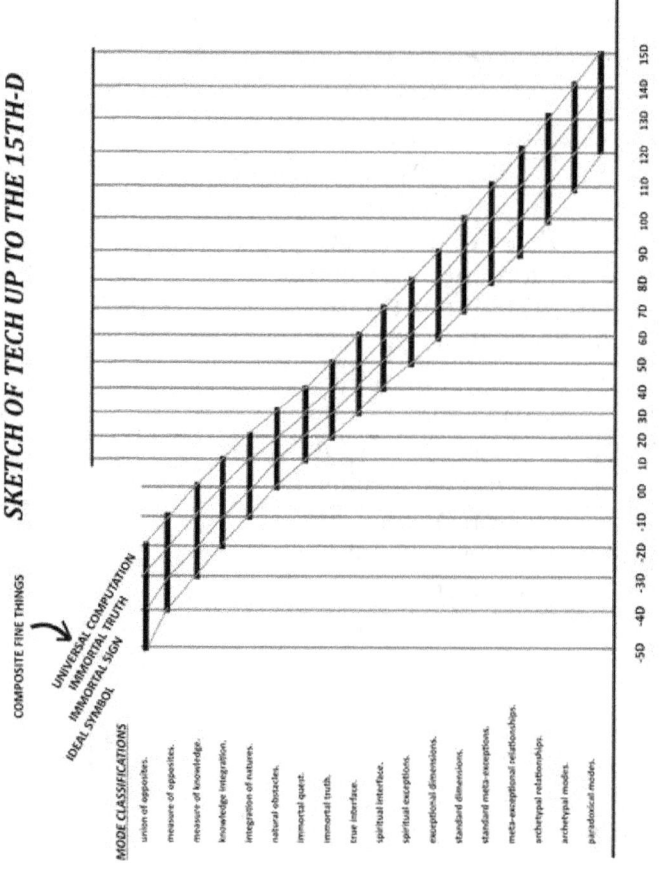

ROUGH TRANSLATIONS OF COMPOSITE IDEAS FOR TECHNOLOGY:

Ideal Symbol in -5-D: union of opposites.

Ideal Symbol in -4-D: measure of opposites.

Ideal Symbol in -3-D: measure of knowledge.

Ideal Symbol in -2-D: knowledge integration.

Ideal Symbol in -1-D: integration of natures.

Ideal Symbol in 0-D: natural obstacles.

Ideal Symbol in 1-D: immortal quest.

Ideal Symbol in 2-D: immortal truth.

Ideal Symbol in 3-D: true interface.

Ideal Symbol in 4-D: spiritual interface.

Ideal Symbol in 5-D: spiritual exceptions.

Ideal Symbol in 6-D: exceptional dimensions.

Ideal Symbol in 7-D: standard dimensions.

Ideal Symbols in 8-D: standard meta-exceptions.

Ideal Symbols in 9-D: meta-exceptional relationships.

Ideal Symbols in 10-D: archetypal relationships.

Ideal Symbols in 11-D: archetypal modes.

Ideal Symbols in 12-D: paradoxical modes.

Immortal Sign in -4-D: union of opposites.

Immortal Sign in -3-D: measure of opposites.

Immortal Sign in -2-D: measure of knowledge.

Immortal Sign in -1-D: knowledge integration.

Immortal Sign in 0-D: integration of natures.

Immortal Sign in 1-D: natural obstacles.

Immortal Sign in 2-D: immortal quest.

Immortal Sign in 3-D: immortal truth.

Immortal Sign in 4-D: true interface.

Immortal Sign in 5-D: spiritual interface.

Immortal Sign in 6-D: spiritual exceptions.

Immortal Sign in 7-D: exceptional dimensions.

Immortal Sign in 8-D: standard dimensions.

Immortal Sign in 9-D: standard meta-exceptions.

Immortal Sign in 10-D: meta-exceptional relationships.

Immortal Sign in 11-D: archetypal relationships.

Immortal Sign in 12-D: archetypal modes.

Immortal Sign in 13-D: paradoxical modes.

Immortal Truth in -3-D: union of opposites.

Immortal Truth in -2-D: measure of opposites.

Immortal Truth in -1-D: measure of knowledge.

Immortal Truth in 0-D: knowledge integration.

Immortal Truth in 1-D: integration of natures.

Immortal Truth in 2-D: natural obstacles.

Immortal Truth in 3-D: immortal quest.

Immortal Truth in 4-D: immortal truth.

Immortal Truth in 5-D: true interface.

Immortal Truth in 6-D: spiritual interface.

Immortal Truth in 7-D: spiritual exceptions.

Immortal Truth in 8-D: exceptional dimensions.

Immortal Truth in 9-D: standard dimensions.

Immortal Truth in 10-D: standard meta-exceptions.

Immortal Truth in 11-D: meta-exceptional relationships.

Immortal Truth in 12-D: archetypal relationships.

Immortal Truth in 13-D: archetypal modes.

Immortal Truth in 14-D: paradoxical modes.

Universal Computation in -2-D: union of opposites.

Universal Computation in -1-D: measure of opposites.

Universal Computation in 0-D: measure of knowledge.

Universal Computation in 1-D: knowledge integration.

Universal Computation in 2-D: integration of natures.

Universal Computation in 3-D: natural obstacles.

Universal Computation in 4-D: immortal quest.

Universal Computation in 5-D: immortal truth.

Universal Computation in 6-D: true interface.

Universal Computation in 7-D: spiritual interface.

Universal Computation in 8-D: spiritual exceptions.

Universal Computation in 9-D: exceptional dimensions.

Universal Computation in 10-D: standard dimensions.

Universal Computation in 11-D: standard meta-exceptions.

Universal Computation in 12-D: meta-exceptional relationships.

Universal Computation in 13-D: archetypal relationships.

Universal Computation in 14-D: archetypal modes.

Universal Computation in 15-D: paradoxical modes.

So, I would say in knowledge at least, the singularity is already upon us!

(SINGULARITY INVENTIONS NOTES)

Some exceptional systems can be measured in a score usually below one, given by the formula:

$(D \wedge Results) - (Verbs - 1)$

Where Verbs equals the number of categories, or otherwise D plus 5.

You may find this formula will not give a value higher than 0.8 or so for things that cannot be used effectively in knowledge.

It is possible to have a bad system that fits the right numbers, but if you have good numbers, they are related to some useful systems.

I call this the Over-Unity Formula for TOE's.

Some past scores: Math (0.10), Objective Knowledge (1), General Solution to Problems (1), Magic Formula (4), Theory of Everything (9), Formula for Souls of Literature (11), higher formulas sometimes 50s, 60s, millions.

(...)

§2: MATCHIK INVENTIONS

BASIC NEED

Repeat "September 3, 2016" and your basic needs will be met.

BLUE PILL, AVOID TAKING

Repeat "Blue pills are not sublime they are not divine".

BRAIN SUGAR

September 2, 1978 Mindful sugar.

EASY CLASSES

Apple juice.

EVOLUTION

Repeat "15 April, 2018"

GOOD GRADES

Apple cider.

HAPPINESS

Repeat "June 16, 2017" and you will find happiness.

IMMORTALITY

Jiaogulan: possibly causes eyes to eventually go black, see in dark, see negative colors when wearing sunglasses, become immortal. Jiaogulan also increases strength, stamina, and mental performance.

The best we can do is think of a goddess and then make high technology unless we're immortal. Maybe all technology is vibrators. I was better off thinking of women as opposed to technology if I wanted to be straight. Maybe this is why programmers seem to get fat.

IMMORTALITY, UNDERSTANDING

July 23, 2015. Understanding immortality and time-travel.

LUCK

Say or attract real time travelers.

LUXURY

Boring knowledge.

MADNESS, CURE

Repeat May 27, 2019

METAPHYSICS

STEP 1: A sufficient number of antipodes might describe the universe in many dimensions, because antipodes can measure two sides of infinity, and a certain number of antipodes should measure a certain number of different measurements concerning everything (or infinite infinity).

STEP 2: Not just any set of opposites will do. One must choose opposites which are actually opposite, and one must also find a way to have perfectly good reasons to privilege the data over other data. This will require actual privileged knowledge, which means exponentially efficient knowledge, as knowledge which is not exponentially efficient will just look like more data of the ordinary kind. Efficient data could at least be technologically useful. And we know things like writing and reading are technology, so exponentially efficient knowledge could mean better writing and better reading, which in principle means better knowledge.

STEP 3: Incoherence just means skills, but skills might not provide a picture of everything. Paradoxes aren't usually solved, so paradoxes should be solved in this system. Irrationality is a bit okay in rare cases, so emotional perspectives should be possible in this system, we can use qualia like it is empirical. Neutral categories are okay because they don't compete with anything, neutral can still be okay. Universals are okay if they really are universals but we can't assume they are because they might not provide us information or clues on everything, not every universal is a theory of everything. Informal methods aren't very useful compared to real knowledge but they could be useful in a classroom. Relativism kind of has a point, but what if you relativize relativism? You can always do that, so relativism doesn't really eliminate knowledge it just makes it look a bit smaller. Quantum? Well, quantum would be explained in a theory of everything, a theory of everything is a different system. It's possible metaphysics is not science-y enough to require an answer to quantum properties. Nonsense systems? Nonsense just involves individual polar opposites which were already included in the system. Impossibility? Impossible systems could have trump, but they involve breaking the rules usually. Breaking the rules could be possible, but it's like changing everything around. Changing the rules isn't acceptable all the time, or probably doesn't have to be. And there aren't a lot of genuine impossible systems that do anything useful. Plus, if one thing is impossible, I could break the rules and call a different system impossible too.

STEP 4: If polar opposites are opposed diagonally this means that opposites are appropriately opposed over the longest possible distance. It turns out this creates exponential efficiency, because the only possible combinations involve swapping the diagonal.

STEPS 5 and 6: If we read myriads of examples involving real opposites like abstract versus material and false versus actual with some clarification we get combinations like: false abstractions make actual materialism and false materials make actual abstractions, which express reasonably accurate knowledge without much tweaking. Some of the cases are even somewhat complex, like fire gives water away. It could theoretically explain infinite properties of nature.

STEP 7: Even though this system works, it doesn't mean that it is a divine law that it works, it's just a human conjecture that holds little weight, but it might be useful to philosophers. It's important that it holds little weight, because if we think like a scientist we must assume it could be wrong.

MULTIVERSE

2021-11-23 It was found there are 25 fundamental ways to change universes:

0. Forget impossibility.

1. Adopt incoherence.

2. Stop growing.

3. Get stopped.

4. To not be naturally mathematical.

5. To stop being rare.

6. To lose your identity.

7. To be unemotional.

8. To be unimportant to humans.

9. To be dysfunctional.

10. To be un-intellectual.

11. To be easy.

12. To be coherent.

13. To be primitive or posthuman.

14. To be unfortunate.

15. To be unclear.

16. To escape time.

17. To be healed.

18. To have effects that disappear.

19. To not communicate.

20. To gain rationality.

21. To go towards the center.

22. To be ugly.

23. To not have a special advantage.

24. To significantly lack opportunity (though this is bad).

OPPORTUNITY

Antidimensions.

PERPETUAL MOTION ELEVATOR

Large balloon attached to small balloon by rod. A second rod specifically weighted and rotated with about a 3:2 ratio to rotation point will be able to cause the whole apparatus to fall when rotated towards small balloon, and rise when rotated towards large balloon, using a drag effect.

("SIMPLE") PERPETUAL MOTION

<u>In your opinion, what is the most mysterious phenomenon of physics?</u>

SURVIVAL, TENACIOUS

Repeat Oct 12, 2018.

THEORY OF EVERYTHING

STEP 1: It probably concerns both the abstract and the concrete.

STEP 2: The abstract means 'knowledge' and the concrete means 'energy'.

STEP 3: It turns out, 'knowledge' means negative differences in an equation similar to 2 (n +/- 1), and 'energy' means perpetual motion in the same equation.

STEP 4: Specifically, both work the same way in the formula Results >= Eff + Diff.

STEP 5: Following an appropriate procedure, the formula works to describe many things, though we as a human species may not be done testing for quite a while.

STEP 6: Specifically, abstraction represents closed sets with efficiency < 1, and energy represents open systems with efficiency >= 1. Thus the formula deserves this as a notation: Results >= Eff* + Diff.

STEP 7: Maybe there are other formulas in other universes, or maybe other universes are the same as saying other interpretations, or other 'readings' of the same formula or something similar, or something describing everything from a different angle. For example, we could write: Efficiency = Results - Difference, OR Difference = Results - Efficiency. These formulas could also be equal and even other formulas as well. There could be a whole group of universal formulas.

TRIP, TYPICAL TRIP

Repeat September 14, 2021

UNIVERSALS

Coherence.

WEIGHT LOSS

Feel underneath far right rib: reduces fat in gut.

NEW

Modulus:

- Of exclusive contents [+]
- Of repetition of infinity [=]
- Meaning [*]

One degree of everything, infinitely repeated, with something meaningful.

§3: TRACTATUS OF STRATEGIC KNOWLEDGE

(Some items removed)

AVOIDING EXTINCTION

Don't do something just for demonstration purposes.

...

BLACK SWANS, HISTORY OF

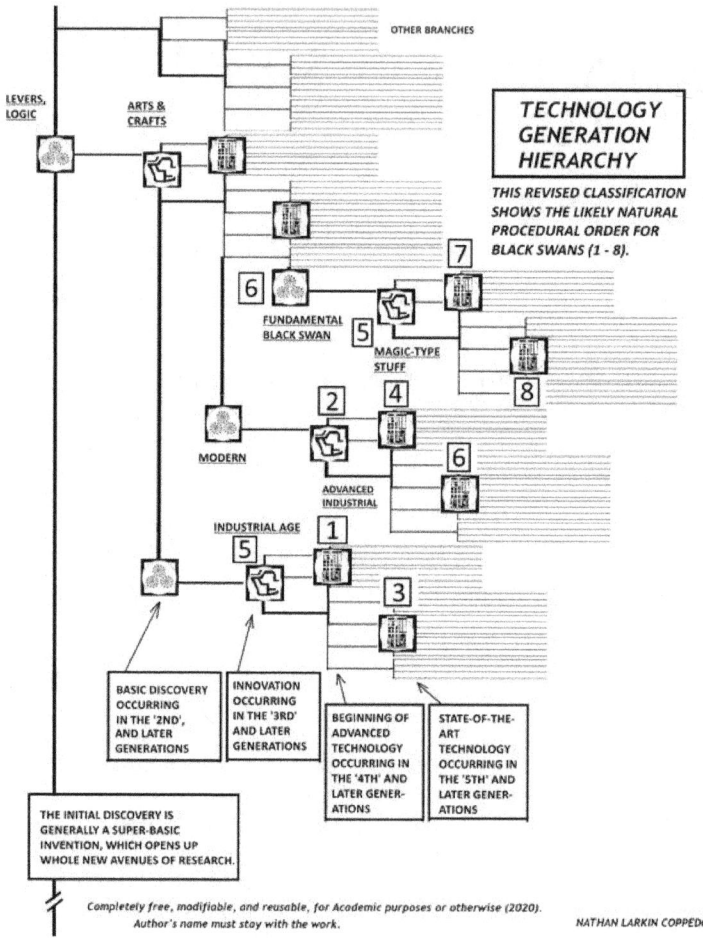

—Classification of Black Swans

BUOY-LEVER

A version of the buoyancy lever, the basis for a likely flying machine principle (an earlier experiment was run on July 16, 2020):

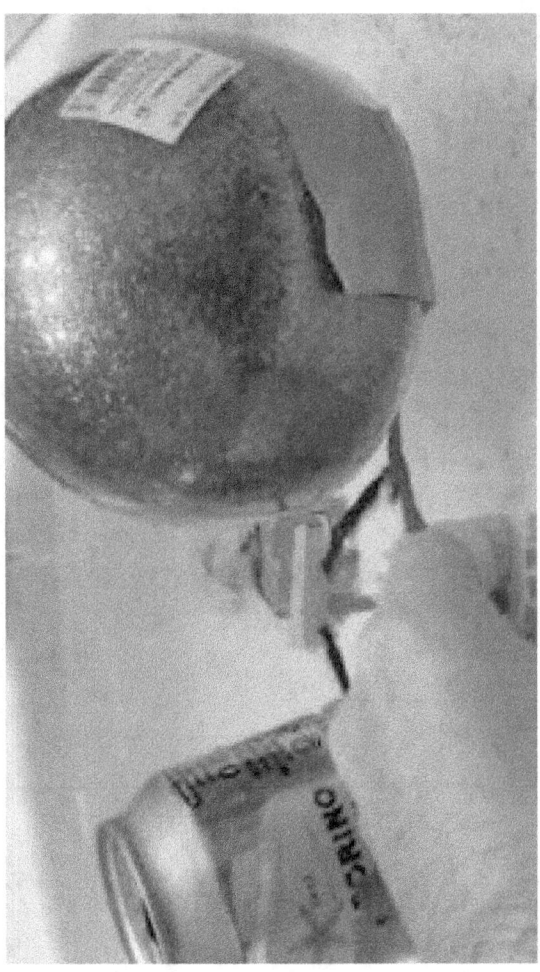

COHERENCE, COMPATIBLE WITH SCIENCE

Original Scientific Method:

- Idea.
- Organize.
- Apply Theory.
- Investigate range.

Coherence:

- Investigate range.
- Apply Theory.
- Organize.
- Idea.

Thus, coherence becomes the mirror-image of the scientific method!

—The Iteration of Science

COMPUTATION ADVANCES

(2-METHOD) SIMPLE METHOD FOR SOLVING COMPLEX PROBLEMS

In 2014 I came up with the formula: the abstract polar opposite of every necessary word in the BEST definition of the problem, arranged in the same order as the corresponding original words.

1. Problem "A B C D" [Any number of parts].

2. Solution "Opp A Opp B Opp C Opp D" [The same number of parts as (1)].

Problems are solved by solutions, solutions are solved by problems.

—What is a general method for solving all paradoxes?

...

(4-METHOD) CHEATS FOR CONVENTIONAL COMPUTING

I dare you to provide proof of these formulas. They are very wide-ranging. By soul here, what is meant is the essential nature of information concerning something, sometimes with some appropriate categorical association like an intelligent concept web (no major emphasis on that).

GIVEN QUESTION (A) is C: B-D SOUL is BCAD or DCAB

GIVEN QUESTION (B) is D: C-A SOUL is CDBA or ADBC

GIVEN QUESTION (C) is A: D-B SOUL is DACB or BACD

GIVEN QUESTION (D) is B: A-C SOUL is ABDC or CBDA

—<u>Universe DNA</u> (...)

...

(8-METHOD) PERHAPS FOR SOME QUANTUM APPLICATIONS

These may be the objective rules of causality.

VALID CAUSAL DEDUCTIONS (In order of decreasing strength and increasing simplicity)

(1) conclusion D follows result of B concluding results of C can be inferred from A

(2) conclusion D follows result of B concluding results of A can be inferred from C

(3) conclusion B follows result of D concluding result of C can be inferred from A

(4) conclusion B follows result of D concluding result of A can be inferred from C

(5) conclusion C follows result of A concluding result of D can be inferred from B

(6) conclusion C follows result of A concluding result of B can be inferred from D

(7) conclusion A follows result of C concluding result of D can be inferred from B

(8) conclusion A follows result of C concluding result of B can be inferred from D

COMPUTATION STRATEGY

External Computer Brains

- The thinking city (Hyper-Cubism).
- Extendable brain (Modular citizenship, modular consciousness)
- External consciousness as replacement for A.I. consciousness.
- Instinctual media-memory.
- Language-based drugs.

Mathematical Over-Unity

Using electricity mathematically with the principle of exponential efficiency might result in a mathematical perpetual motion machine. (2021–07-08)

...

CORRELATIONS, STRONG

Fancy thing:

Ex: Perpetual motion machines.

(COMPARED WITH)

Substance of 'greater class does':

[For ex, Factories] Plural if possible.

COMBINE FIRST WITH SECOND AS BEST YOU CAN, PLACING THE FIRST FIRST.

...

CREATIVITY STRATEGIES

[1]

STRATEGY 1: Think of new categories.

STRATEGY 2: Build on earlier work, or combine categories.

STRATEGY 3: Artificially invent a new function, then develop it.

STRATEGY 4: Develop, evolve, iterate an earlier movement.

[2]

1. If acute brilliance and brilliant performance is expected, apply creativity and hard work, and let others steal it if they know how.

2. If the problem is deemed impossible, change the medium from math or science to philosophy, and solve the problem generally rather than within time and space. This can simplify, and scientists seek simplified general solutions anyway.

3. If creativity is required, work within art if logic is required. Work within language if innovation is

required. If science is required, then view things progressively, historically, and reductively, looking only for new angles. If there seem to be no new angles on science, apply heavy critique and look for new technologies unless existing technologies were not yet successful. If the current paradigm is complex, look for simple impossible technologies. If the current paradigm is simple, look for complex impossible technologies. Some of this work may already be done by Nathan.

4. If the problem is a matter of unknowns, reverse the largest recent paradigm and apply simplicity or complexity, whichever is least dominant.

[3]

1 IF YOU HAVE A MAGIC WAND, CAN YOU THINK OF SOMETHING NEW TO CAST AS A SPELL?

2 CAN YOU THINK OF A NEW ART MOVEMENT?

3 CAN YOU THINK OF SOMETHING YOU WOULD LIKE TO INVENT?

4 IS THERE AN EXISTING EXAMPLE YOU HAVE IN MIND?

5 IMAGINE YOU ARE A GENIUS, HOW WOULD YOU MODIFY THE EARLIER INVENTION?

Keep in mind one person may have a finite number of inventions, perhaps up to 20 major inventions and their variations, depending on how great the inventions are. For the purposes of discovery one must assume infinite to get anywhere significant.

[4]

Paradigm 1 is if someone is bright, like Aristotle.

Paradigm 2 is if logic becomes the established idea.

Paradigm 3 is if there is a new and better logic.

Paradigm 4 is if the new logic creates a larger paradigm.

—<u>Which ones are the most rational activities that a human being carries out, and why is that?</u> (Logical version of the Golden Age Formula)

—<u>Black Swan Studies</u> (...)

...

DETECTING HIDDEN ROOMS

Elements + Sets - Number of Known Rooms

DETECTING HIDDEN STRUCTURES

D (Logical Efficiency) + D (Differences, or Contradictions, or Sets). <u>Hidden Structure Theory</u> (...)

DOUBLE-LEVER

Key discovery recorded by Nathan Coppedge February 6, 2021:

Keep in mind, this may be the best antigravity device ever created. Depending on the size of the apparatus, the heavy weight could be lifted to very high altitudes merely by rotating a very thin, lightweight lever. This is the ultimate of efficiency for causing heavy weights to be lifted at very low energy cost.

...

ELEVATOR, PERPETUAL MOTION

An 'elevator' is set on a very slightly upward-inclined track that is slotted, with a very long lever passing through, which can be in various ratios. If we assume the elevator has a mass of 1 and the long end of the lever has an unweighted mass advantage of 1, the counterweight's mass will vary from (the max opposing leverage X the gradient + 1) to (the min opposing leverage + 1). Since the gradient should be about 0.6 or 0.7, and the device functions much like a balance, if it is low friction enough with the correct ratios, and the 'elevator' is supported by the track on its upward motion, but deflected sideways and only supported by the lever on the downwards motion, then this device appears to work.

Perpetual Motion Elevator (...)

HORIZONTAL ADVANTAGE (MOVEMENT)

Below: In the Ball-and-the-Wall experiment (January 2014 day of month unknown): if momentum equals 1/2 mass X distance expressed in mass like a wheel, why does this ball roll well over 4X the distance it fell? We're not calling this occult.

INVISIBILITY

If OU = 0.96875 , "TRUE".

LIGHTER WEIGHT CAN LIFT HEAVIER WEIGHT

1/2 m x d experiment, a result of analyzing the so-called Superman Fallacy, a stick of 65% mass was able to lift a longer stick which had 100% mass, upward along a smooth surface, when the 65% mass acted through a string which was pulleyed (October 14, 2019).

MATH, FIND RELEVANT NUMBERS

GENIUS TIP: 1 /(((Min Eff+1 - (Max Eff / 2)+1)/ Efficiency) + 1)

SPECIAL VALUE THEORY: Special Value = [1 (Eff) + 0.5 (Diff)] - D. [January 12, 2021]. Special Value Theory (...)

MACHINES, SIMPLE

Some simple machines: Prop, Pry Bar, Lever, Screw, Wheel and axle, Pulley, Wedge, Reverse wedge, Ramp, Pendulum, Rope / string, Funnel, Balance, Counterweight, Hinge, Swivel/ Pivot, Turntable, Clamp, Drawbridge, Latch (hasp lock), Bow, Blunt object, Blade, Bearings, Deadly Trap, Ratchet, Tent-Pole, Crank, Winding mechanism (winch), Blocking mechanism (bevel), Straw, Bellows, Extendor bars, Buoy, Bridge, Balloon, Kite / Glider, Zip-line, Sundial, Prism, Mirror, Art, Camouflage, Pen and Ink, Reservoir / Shunt / Feeder / Thrower / Siphon.

...

OBJECTIVE KNOWLEDGE, MATHEMATICS OF

KNOWLEDGE

2D 1 result 2 categories

2D 2 results 4 categories

2D 4 results 8 categories

2D 6 results 12 categories

2D 8 results 16 categories

4 D 1 result 2 categories, etc.

8 D 6 results 8 categories

8 D 12 results 16 categories

8 D 15 results 20 categories

16 D 5 results 8 categories

16 D 10 results 16 categories

16 D 15 results 24 categories

...

OBJECTIVE KNOWLEDGE, MEASURING POTENTIAL

Some exceptional systems can be measured in a score usually below one, given by the formula:

(D ^ Results) - (Verbs - 1)

Where Verbs equals the number of categories, or otherwise D plus 5.

You may find this formula will not give a value higher than 0.8 or so for things that cannot be used effectively in knowledge.

...

PERPETUAL MOTION, STATISTICS

- Min Heavier Mass = (Max Lvg / 2) + 1

- Max Heavier Mass = Min Lvg + 1

- Min Lvg = Max Heavier Mass - 1

- Max Lvg = (Min Heavier Mass - 1) X 2

- Over-Unity = Heavier Mass Rng / Lvg Ratio + 1 X 100 (%)

- Smaller Mass = 1X

- PM Cars Extra Mass < OU - 100%

- Flying Vehicles Extra Mass < OU - 200%

- Flying does not work when Lvg Rng >= 1/2 max leverage.

- Flying Machines Window = Max Larger Buoyancy - Min Larger Buoyancy

- Flying Max Larger Buoyancy = (Min Lvg) additional mass cancels with 1 unit buoyancy

- Flying Min Larger Buoyancy = (Max Lvg / 2) additional mass cancels with 1 unit buoyancy

- Flying OU = Larger Buoyancy Range / Leverage Ratio + 1 * 100 (%) + 100 for buoyancy.

- Flying Smaller Buoyancy = 1X

- Secret of perpetual motion flying machines: <u>Improved Balancing Balloons Theory</u>

- Planetoids: Estimated < (Phi / 2 + 1 * 100 =) 180.9% OU,

- Max sustainable mass resistance to Earth perpetual motion = <<0.809 X distance (Earth diameters). With an estimate saying Earth's max output is about 110% with rotation.

- The Max Min required distance to resist the Sun with PMMs is 4726 AU assuming 110% OU

- Perpetual motion holds the key to the material world.

- Flying machines hold the key to the universe.

PSYCHIC PREDICTIONS

What is it? Something argues for it. What created the argument? What does it sound like when you say who created it is not what it was?

—Abbreviated Secrets of Prediction

...

RESEARCH ANYTHING

Title of book = '[quality of X] [opp qualifier]'

Information on the book = 'If you [X] [qualifier] subject of X and qualifier [opp X clarified]'

—(Formula by N Coppedge from 2016)

SUPER-STRATEGIES (CULTURAL STRATEGIES)

Chinese:

- Martialing the arts.
- Overwhelming forces.

Phoenicians

- Overwhelming forces.
- Religious traditions.

Egyptians:

- Religious traditions.
- Appealing culture.

Greeks:

- Appealing culture.
- Enslaving others' cultures.

Romans:

- Enslaving others' cultures.
- Consolidating power.

French:

- Consolidating power.
- Robbing the people.

British:

- Robbing the people.
- Industrialization.

Americans:

- Industrialization.
- Mass production.

Rich culture?:

- Mass production
- Perpetual motion?

...

THE THEORY OF EVERYTHING

The Theory of Everything (theory of anything, June 26, 2019).

Total Results >= Total Efficiency* + Total Difference

*Where difference = results - efficiency, and where efficiency sums to < 1 if topic is acted on, and sums to > 1 if topic is acting. Since the Efficiency and Difference are correlated in the Result, the system does not commit circular reasoning if two variables are given empirically or rationally.

Step-by-Step Process:

1. Is the phenomena active or instead passive? Is it acting or instead being acted on? If it is active, efficiency will be greater than 1, usually a whole number, if it is passive, efficiency will be made of multiple parts adding up to 1.
2. Now that the efficiency can be determined, what is set 0? What is the number of things acting or acted on, as this is set 0? You may simply use negative, 0, finite, infinite, etc or you may give a specific finite value.

3. Now find the Difference between the Set 0 value and the efficiency, and add it on the end of the equation as the specific difference (like a constant) for that exact problem.
4. Now, assuming the formula makes sense (as it is expressing limits, Set 0 will equal total efficiency + difference), translate the meaning of the efficiency in terms of the difference vis Set 0 to get a Theory of the Subject.

...

Theory of Everything on One Page

TOE: Results >= Efficiency + Difference | Anti-Theory <= Difference - Efficiency

Efficiency >= Results – Difference | Anti-Efficiency <= Difference - Results

Difference >= Results – Efficiency | Anti-Difference <= Efficiency - Results

Max OU <= (((Min Eff+Diff) - ((Max Eff/2) + Diff)) / ((Max Eff + Min Eff) /2)+Diff)

Anti-Energy >= 1 - (D + Difference)

Ideal Elements = D + 2 | Un-Ideal Elements = - 2 - D

Ideal Principle = Negative [(D - 1) ^ 2] | Un-Ideal Principle = [Sq rt 2 (1 - D)]

Basic Meaning = 5/32 proportion | Incomplete Meaning = 160 (constant)

Math Form = 0.10 X Absoluteness | Above Math = Qualities X 10

Languages = ifs(Dimensions<=0,"0",Dimensions<4,(1.585*1.09^(Dimensions-2)),Dimensions=4,(1.585*1.09^(Dimensions-

2)+1.09^(Dimensions-3)),Dimensions>4,(1.585*1.09^(Dimensions-2+N(1)+1.09^(Dimensions-3+N(1))))) [Thought to be effective in all higher dimensions]

Anti-Languages = [1.09 rt of (2 minus D)] / 1.585

2 / Avg Speed = Observed (Theoretical) | [Sq rt of 0.5 (Time)] / Avg Speed = Detected | The first observer aims to refute. The second observer acts passively. The first particle responds quickly. The last particle is a slave.

Dimensions - Antiforces = # Forces | Dimensions - Forces = # Antiforces

Dimensions = Forces + Antiforces.

$(D^2 +1)^2$ = Grand Theory number

Disintegral = - (Difference – Efficiency) | Special Value = [1 (Efficiency) + 0.5 (Difference)] - D

Anti Disintegral = Efficiency - Difference = Antitheory

Antivalue Theory = (Dimensions+(Difference / 0.5 not a mistake) - Efficiency)

Simple disintegral = 1 / Antivalue | Nodes (simpleforms) = Minus Antivalue

Verbs = Difference + 5

OU Formula for TOEs = [(Dimensions ^ Results) - Verbs - 1]

General Possibility = |2 / D - (results / (OU + ((D ^ Results) - 1)))|

Exponential Efficiency = Efficiency - Difference

Coherence = Ifs(Results<=0,"Incoherent",Results<1, "Truth", Results=1, 1, Results<=2, Results, Results<=(Results-1+ N(+2)), - (Efficiency - Difference), Results <=(Results+ N(+2)) , 0)

What's needed in an abstract system is -2 (Avg Eff)

What's needed in a physical system is 0.5 (Avg Diff)

—<u>Theory of Everything on One Page 2021–05–04</u> (...)

...

THEORY OF LANGUAGE, POST-UNIFIED

<u>General Proof Regarding Universal Language</u> (...)

Using the 10 categories of the Characteristica Universalis and their opposites:

If something were not a category, it would have nothing.

If something could not stand, it would have a powerful base.

If something could not spread out, it would have length.

If something were incapable, it would swim in a soup.

If something were not alive, it could still be classified.

If something were not a system, it could still be called one.

If something had no substance, it would not exist.

If something was not abstract, it would exist.

If something was not organized, it would be a part of nature.

If something were not flagged, it would be noted.

Thus, everything that speaks a language must have one of these things, the first or the second.

And, so, everything that is missing one language is speaking

- In the broadest sense, if we do not adhere to one modality, another modality applies.

...

TRANSPORTATION

Anachronistic Ground Transport

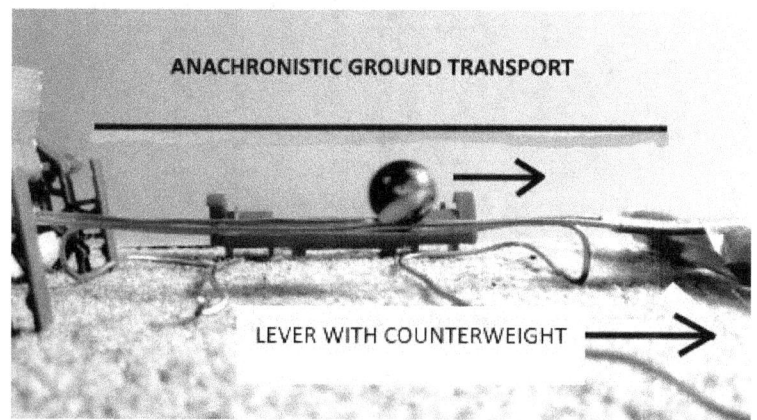

- First Known Theory: August 4, 2020.
- First partial evidence: August 4, 2020.
- Preliminary Theories of Perpetual Motion Trains: December 17, 2018.

- Based On: 1st Fully Provable (July 12, 2016).
- Rating: Avg < 175%
- Leverage: 1.5 to 0.5 :1
- Counterweight Mass: > 1.75X to < 2.5X
- Note 1: It is noted the Min counterweight mass comes from the Max leverage /2 as usual.
- Note 2: However, the Max counterweight mass is not taken from the Min since first application happens at max leverage. This explains the 175% over-unity.

Automatic Rowing

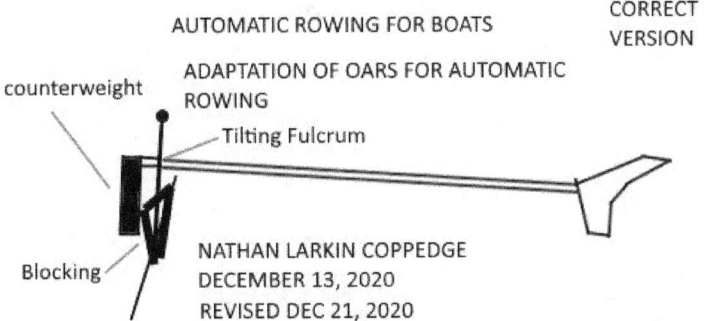

Tilting the fulcrum forward creates forward movement.
Tilting the fulcrum backward returns the oar.

Blocking may be changed to allow return at higher altitude.

Effective Rowing: Improvement:

- Flanges would be fitted to the poles.
- Rowing would take place by shifting the lever backwards all the way, causing the pole to shoot upwards.
- Additional smaller back-positioned counterweights would lift the fulcrum upwards slightly as the lever is shifted backwards.
- The forward motion of the lever would drop the flanges simultaneously forwards with limited resistance due to the medium counterweight and allow the oar to drag a long distance backwards, due to the placement of the net and motion traveled during the drop.

Self-Moving Planets

WARNING: Planetary transportation if it occurs may result in the death of all species.

"MOBILE SPHERE" PERPETUAL MOTION PLANET

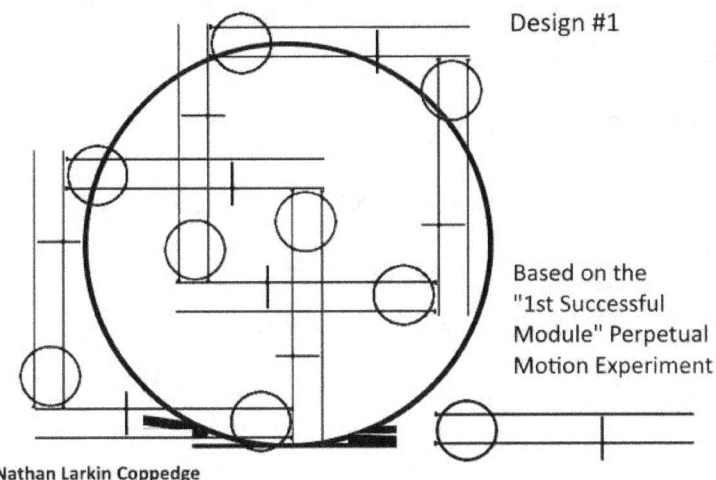

Design #1

Based on the "1st Successful Module" Perpetual Motion Experiment

Nathan Larkin Coppedge

Date of Invention: Nov 30, 2020
Rating: <150% conventional Over-Unity minus mass.
Leverage in each machine: 1:1 (judging by counterweight distance)
Counterweight Mass: >1.5X to <2X ball (assumes 1X additional weight in long end).

Equation: Assuming ball = 1 with variable application, and long end has additional 1 constant application, and counterweight located on shorter end, and counterweight is designed to direct ball on opposite end up slight supporting incline before ball applies leverage,

Unified Counterweight Mass Formula = Min Lvg + 1 > (Max Lvg / 2) + 1.

Date of Invention: November 30, 2020

Precedents: "Starwars" military satellite project, Starwars franchise.

Machine Rating: < 150% Conventional Over-Unity

Machine Leverage: 1:1

Machine Counterweight Mass: >1.5X <2X

Planetoids: Phi / 2 + 1 * 100 = < 130.9% Conventional Over-Unity

...

UNIVERSAL TECHNOLOGY

Based on immortality formula, possibly TOE category combine with category +2:

- Math + TOE.
- Wish + Perpetual Motion.
- TOE + Elements.
- Perpetual motion + Meaning.
- Elements + Function.
- Meaning + Energy.
- Function + Variation.
- Energy + Language.
- Variation + Psychic.
- Language + Organization.
- Psychic + Species.
- Organization + Set.
- Species + Resources.
- Set + Sufficiency.
- Resources + Math.
- Sufficiency + Wish.

UNNATURAL TORQUE

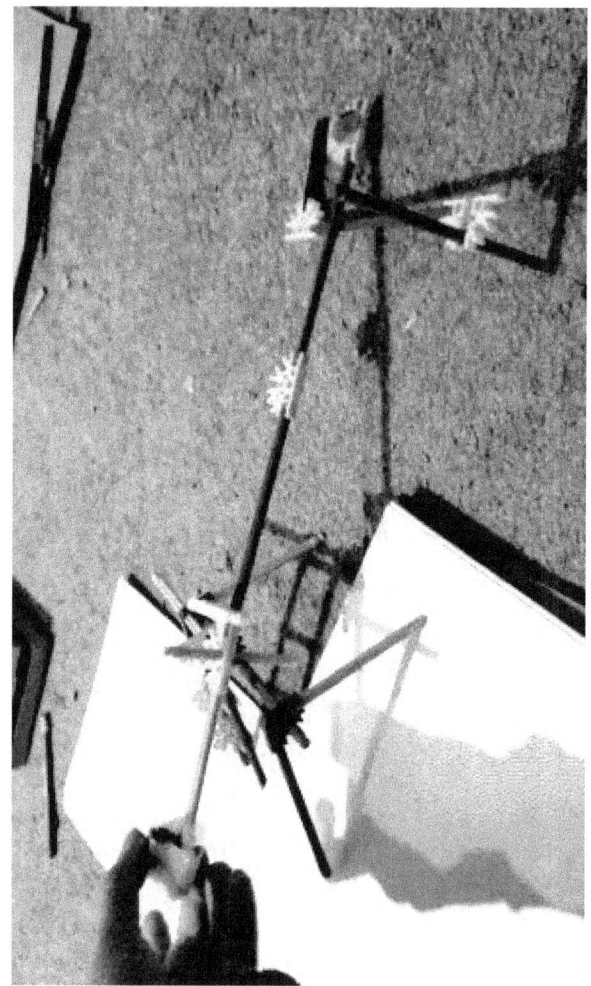

...

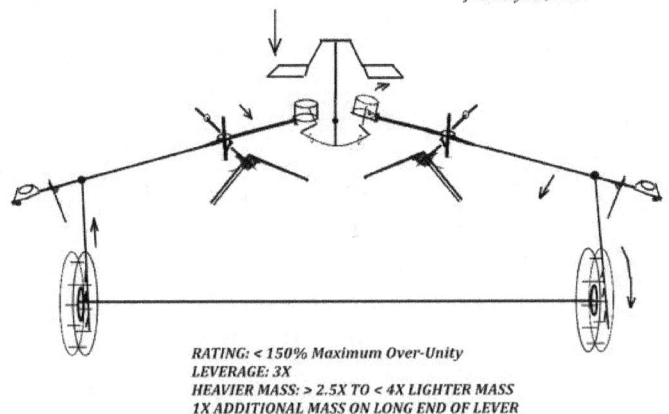

ABOVE: A 'car-wagon' concept powered mostly by the Unnatural Torque.

...

WISDOM

1. Don't assume too much, but you will only survive if you are a little like a devil.

2. Bad devils are often punished.

3. If you are stupid, you should try to be smart.

4. If you are smart, you should try to have fun.

5. Don't waste money unless you're priceless.

6. Try to be wiser than a fool.

7. Don't go out of your way to find something hidden, you will be played false.

8. Some things that cannot be attained by intelligence can be attained by hard work.

9. You cannot always be smartest, but you can try looking smartest.

10. If you have failed at one thing, it can be better to try succeeding at another.

...

§4: SINGULARITY SPELLS

0 AD

Summa Arcana: Black Chameleon Wish, Great Theory-Puzzle Ironically Cursed

...

MODERN AGE

Is like Bertrand (First Name). Desiring evolution.

1907

Assumèd: a thought of Euler about being made of shit brings people happiness when they cannot process things.

1978

September 2, 1978 Mindful sugar.

1982 (Year of the Meaning of Life)

October 23, 1982 Very spooky, meaningful genes. Actually smart. (In Nathan's case a kind of 'Devil', symbolized by reincarnation number two fixed as long as he stays in the same matching set of worlds, consistent since 7 million BC on a 5 and 7 world regular cyclical rotation).

1984 (Year of God's Vanity)

October 25, 1984 Being able to do everything. Finds things obvious.

1985

Everyone becomes more conscious and suffers more.

1989

Intellectual conspiracies.

1991

For some reason or other or no reason at all, intelligence doesn't matter in this year.

1992

This year people feel they are given an offer they can't refuse, like playing Russian Roulette.

1993

This year everyone could be exceptional, even more exceptional than normal.

1994

This year everyone was smarter, smarter than normal, better at school.

1995

This year everything was about one person.

1996

This year everything was realistic and believable.

1997

This year everyone looked to the future.

1998

This year everyone was introduced to advanced ideas.

1999 (The Year of the Devil Reversed)

2000 (Year of Advanced Toys, for example, TECHNOLOGY-2000)

September 2000 Perpetual motion can be invented.

2001 (Year of Good Cartoons, Year of Modern Poetry)

2002 (Year of Chaos)

2003 (Year of the Father, Year of Divine Aspirations)

2004 (Year of Half Truths, Year of First Discovery)

2005 (Year of Design & Pornography)

2006 (Year of Greatness, Year of Strength)

2007 (Year of Ideals, Year of the Perfect Birdie)

2008 (Year of Insanity, Year of Small Wisdom)

2009 (Year of Everything, Year of Doing Everything)

2010 (Year of Judging Fortune, Year of Good Standing)

2011 (Year of Judging Perfection, Year of Good Work, Year of the Writer)

2012 (Year to Consider the End, Year of Resolution)

2013 (Year of Judging Death, Year of Avant-Garde Phliosophy)

FEBRUARY

February 2, 2013: Day of Objective Knowledge, in other words Day of Truth.

2014 (Year of Judging Perfection, Year of Answers to Everything)

At some point this year the formula for the soul was invented or re-discovered.

JANUARY

January 28, 2014: Designs & Theory.

FEBRUARY

February 3, 2014: Day of Problem-Solving.

OCTOBER

October 18, 2014: Highest Artform.

2015 (Year of Judging the Devil, or Year of Bad Talent)

FEBRUARY

February 20, 2015. Sublimism.

JULY

July 23, 2015. Understanding time-travel and immortality.

2016 (Year of Judging Sweetness)

JANUARY

January 8, 2016. Dimensional Art.

SEPTEMBER

Repeat "September 3, 2016" and your basic needs will be met.

NOVEMBER

November 1, 2016. They will hold the key to mastery beyond ordinary examples. <u>The Four Courses of Mastery</u>

November 28, 2016. They will know how to cast spells.

DECEMBER

December 29, 2016. An immortal day.

2017 (Year of Judging Limits)

FEBRUARY

February 2, 2017. Day of High Sorcery.

February 4, 2017. Viruses and chain letters day.

February 7, 2017. Becoming God Day. You decide.

MARCH

March 25, 2017. Everyone born on this day is privileged.

APRIL

April 2, 2017. On this day everyone will be sad, but later they will feel better.

April 8, 2017. Someone is rumored to be about the Programmable Heuristics or about Nathans on this day.

On April 11, 2017 to Cure the Blind in some way utter: "White and Yellow and Red and Orange and Pink and Brown and Purple and Blue." However, this may also work against wisdom and style.

JUNE

Repeat "June 16, 2017" and you will find happiness.

JULY

July 6, 2017. The Day of Medicine. Medical things happen on this day. Someone might be good at treating injuries, and particularly writing like a doctor.

July 7, 2017. Questioning ignorance. <u>The Question of Ignorance</u>

July 26, 2017. Key to a divine body for time-travelers.

AUGUST

Write "August 4, 2017" and you will gain experience. On this day, the less you sacrifice, the more success you gain in the long-term.

Remember "August 10, 2017" and you will be able to time-travel, first back, then forwards.

August 13, 2017. Rational inspirations.

SEPTEMBER

September 5, 2017. The Day of the Artistic Paradise

September 7, 2017. Soulful perspective, represented by one grain of salt.

September 10, 2017. The Day of Eternity for some reason. It keeps going endlessly on, but usually you get to the end. It is associated with perfection or divine reaction-time. Really it is the Day of Divine Reaction-Time.

OCTOBER

October 3, 2017. The Day of Valuable Notes, Hints, Flavors, etc. In other words, The Day of Flavors, if you are a time-traveler.

October 7, 2017. Wisdom and Ignorance. <u>Wisdom and Ignorance</u>

October 25, 2017. Utopianism, Survival, Arcologies, etc. In other words, the Day that Elite Snobs Take Over with a Pretty Good Idea that Doesn't Turn Out Well.

October 28, 2017. Rumor is someone can invent the soul on this date. They may also be good at magic, inventing, and dating.

NOVEMBER

November 25, 2017. Wave Day.

DECEMBER

December 9, 2017. A day for sophistication.

December 13, 2017. A day for scientific souls.

2018 (Year of Judging Madness)

JANUARY

January 2, 2018. High creativity. <u>Creativity Links</u> (...)

January 20, 2018. Someone born on this day can very likely speak secret languages.

FEBRUARY

February 11, 2018. Unavoidable Randomness.

MARCH

March 14, 2018. They will be concerned with the Intermediate. The middle-ground. The possibilities. <u>Intermediate Studies</u> (...)

APRIL

Spell of Evolution: Repeat "15 April, 2018". Someone born on this day is one of the most evolved people.

MAY

May 3, 2018. Day of Propinquity, or just Pink.

JUNE

June 15, 2018. Possibly the Day that God Gets what He Wants.

JULY

July 2, 2018. Whoever is born on this day has an affinity for spiders and knows about perfect internet. Tract of the Ideal Cosm

July 6, 2018. The Good World. The Good World

Late July 2018. May be victims of computer hacking.

July 26, 2018. The Day of the Anti-Room or anteroom.

SEPTEMBER

Perfect Evolution: Things related to "Septembert 2018"... [Not a spelling mistake].

September 5, 2018. This day is devoted to Metaphysics, or sometimes Meta-Physics. It is a special day for both metaphysics and physics.

September 10, 2018. Genius ideas. 10 grains of salt day.

OCTOBER

Oct 12, 2018. They are tenacious survivors. Relevant quote: pain is always worth an interface.

DECEMBER

Dec 13, 2018. Something having to do with a Leviathan.

Dec 16, 2018. People are authentic on this day. For a moment, no one is an actor.

2019 (Year of Judging 'Sun-light')

FEBRUARY

February 22, 2019. Very good at indexing or pointing. The Ideal Index

MARCH

March 23, 2019. Day of brainy viruses.

APRIL

April 19, 2019. Great Inventor Day. Or, great inventor is God to them.

MAY

May 4 - 5, 2019. Possibly the Evening of Shapeshifting if you are a time-traveler. Metaphor = red pill = changing form.

May 26, 2019. Their life will be state-of-the-art.

Cure for Schizophrenia: "May 27, 2019".

May 29, 2019. Poems to save the eyes. Poems to save the soul. Reason by Exercise of the Eyes (if we save the top of our eyes from worms then we're innocent).

JUNE

June 19, 2019. Get what you want: cite unique identity, say: 2.0.1.9.0.6.1.9. ... 2.0.1.9.0.6.1.9.

June 26, 2019. Everything Day. If you are born on this day you likely know everything and have some type of endowment.

2020 (Year of Judging Judgment)

JANUARY

January 3, 2020. Wizard Sports. <u>Wizard Sports</u>

January 24, 2020. Day for Exorcists to be born.

January 28, 2020. It is thought this is the real Rube Goldberg Machine day.

APRIL

Repeat "April 23, 2020 to May 19, 2020" and you will gain more universal genes.

April 27, 2020. The Day of Power. Like omnipotence.

AUGUST

August 18, 2020. Day of Thinking Like Vampires.

August 30, 2020. Day of Moderately Weak Warding. <u>Warding Non Sic</u>

NOVEMBER

November 21, 2020 Day of the Problematic Paradise.

DECEMBER

Repeat "December 19, 2020" big rolls, devillish power, good luck.

2021 (Year of Judging Luck)

FEBRUARY (Fat month)

February 10, 2021 Fat Load of Inventions Day. <u>A Meaningful Directory of Black Swans</u>

February 15, 2021 Universe DNA

February 20, 2021 Someone born on this day will be able to fly, maybe just once.

February 24, 2021 Bigger perspective on Thermo. Correct expansion of the Theory of Everything.

APRIL

April 14, 2021. Rumor it is easier to find Earth on this day. Actually, you can use this same expression any day: "If you're not a philosopher, you must be good at arguing." to help find Earth and gain an existence there. [Antihesis: "This expression was invented on this day."]

MAY

May 23, 2021. The Desired effect is possible.

JULY

July 12, 2021. Moral medicine. <u>Moral Medicine</u>

AUGUST

August 1, 2021. Technology spells. At least they will be realized. Actually it is a terrible gamble. If they are realized, we win. If they are not realized, we lose at least to some extent. Technology Spells

SEPTEMBER

September 14, 2021. Whatever you 'trip' on this day will always be what you 'trip' like: Trips (...)

HYPER-FUTURE

Is like Bertrand (Last Name), something too futuristic happens.

ADDITIONAL NOTES

Dystopianism: Say: "Factor A and Factor B". (Addictive to dystopians)

Principle of Animal Intelligence: Biology would say animals are conscious too, because otherwise humans are only a hair above complete imbecility.

NOTES ON IMMORTALITY, TIME-TRAVEL, TELEPORTATION, TELEKINESIS

NEWS: Jiaogulan probably causes black eyes

- Evidence: I have observed when I take jiaogulan for about six years, I begin to see better in the dark, but I begin to see darker colors in the light. Perhaps sunglasses are for this purpose.

...

Conversation I overheard from a teacher with an Asian girl who had black eyes:

I thought all of your kind were killed.

I'm sorry.

You probably should be killed.

...

Helpful on living to 150 years...

- **DOLPHIN BRAINS**
  - ADVANCEMENT ---> LITERALISM
  - SOPHISTICATION ---> MAYBE A USEFUL TRICK

- **THE COHERENT BRAIN**
  - WHAT OF POWER? ---> IDEA ---> ENCOUNTER
  - IS IT DIVINE?

- **THE YIN-YANG**
  - WE HOPE FOR LATER --->
  - IF NOT EARLIER

- **THE QUICKENING**
  - "Blue pills are not sublime they are not divine"
  - RED PILLS ---> PROMISE ---> PROBLEMS --->
  - CLARITY ---> MAKE A STORY   AMBITIOUS SOLUTIONS
  - ISOLATION ---> THAT IS ALL ---> TRY BEING SAD --->
  - EYES ---> A BAD MEMORY   MAYBE YOU'LL REMEMBER

- **KERNEL NATURE**
  - LIFE ON THE MOUNTAIN --->
  - GEOMANCY ---> MATH ---> SECRET --->
  - EMPEROR'S SOUP ---> GENERALIZE   BUSINESS
  - OLD SONGS --->

- **SIDEREAL SKY**
  - GILT --->
  - IMAGINATION ---> COHERENT ---> OCCULT --->
  - MIMICRY (IMPRESSIONS) ---> PERPETUAL MOTION   REMOVING A PARASITE

Imperative history: Part of the attraction to wheat in Europe during Medieval times was that Jiaogulan was thought to be colored like wheat.

<u>Quantum Immortality</u> solution to this problem involving intelligence preferences and perception of what is beneficial to intelligence.

Current Immortality Plan:

(1) Spark of genius.

(2) To be more ambitious has ultimate worth.

(3) There are paths and there are levels.

(4) There are many avenues in the forest, many passages in the ways of animals.

(5) What if something springs up, divine and new-formed?

(6) Of the many fables of the mind, one of them holds true, to live forever and be immortal.

...

Possible Formula for Immortality based on Improbability Drive:

REDUCE NEGATIVE INFINITE RESULTS TO NEGATIVE FINITE RESULTS

REVERSE INFINITE NEGATIVE EFFICIENCY BY SEEKING FINITE IMPOSSIBLE RESULTS

APPLY STANDARDS

IGNORE RESULTS, MAKE EFFICIENCY INFINITARY USING KNOWLEDGE

MAINTAIN INFINITE EFFICIENCY, APPLY IMPROBABILITY

...

BASICALLY, INFINITE STANDARDS, FINITE OBSTACLES

TIME-TRAVEL METHODS

How to Build a Time-Machine

1.

COPPEDGE'S TIME MACHINE CONCEPT:

Components:

- Local matter deep-scanner.
- Quantum exotic matter time-locator and wide-ranging exotic matter map generator.
- Exotic matter generator / programmer.
- Seats / booths.
- Universe / dimension synchronizing tool / computer, mostly for making sure people and equipment can be translated as exotic matter from inertial reference frames.
- Event initializer / time editor / opportunity locator / computer for locating or else creating commonalities in timelines for the purposes of synchronization.

I'm not sure you could make equipment that way, but you could try.

By the way, I have no direct experience with time-travel equipment except my own mind and perhaps magical people or disguised vehicles.

All of the above is merely an interpretation.

2.

Alternately,

YANG'S ORIGINAL TIME MACHINE CONCEPT:
- Skill with history.
- Ethics.
- Practical advice.
- Knowledge of physics.

3.

TIME CRYSTAL TRANSPORTER

The loop itself has finite energy, but appears to time-travel in a loop while it has that energy... they are real, in some highly special cases... It seems like they could be used to create a time-travel vehicle.

—<u>What kind of energy do time crystals give off?</u>

MAGIC TIME-TRAVEL METHODS

- Gain a position of rank and apply classicism and leverage. This method includes the following sub-methods: 1. To be a god of politics, 2. To be a politician with orthodox methods, 3. To be a politician with advanced technology, 4. To be granted exception of rank.
- Retain a position of justified insignificance, develop a complex value system, and transcend by being tailored out of the system. This method includes the following sub-methods: 1. Ersatz mentality in a position of limited faith, 2. Linguistic ability without efficacy, 3. Subjection to pruning by acting powers, 4. Artificial interface assumption.
- Become a time-travel actor in a justified routine. This method includes the following sub-methods: 1. Be a principal actor under a time-travel politician, 2. Have secret knowledge of a justified-insignificant time traveler, 3. Be technically significant in an insignificant context, 4. Be classically significant in a significant time-travel context.
- Develop a qualified time-travel significance. This method includes

the following sub-methods: 1. Originate knowledge of popular or technical time-travel, 2. Be highly unique in the claim of relating time and travel, 3. Have data which serves as a basis for an investigation of time travel, 4. Have reason to engage in time-travel.
- Slide into time-travel from a related program. This method includes the following sub-methods: 1. Learn teleportation, 2. Build a perpetual motion machine, 3. Found a government or civilization, 4. Be born into an immortal family.
- Attempt a bureaucratic time travel. This method includes the following sub-methods: 1. Master highly specific complexities of an office job, 2. Engage in those highly specific complexities, 3. Become forgotten to time, 4. Find a significance for re-emerging from the details.
- Childhood time travel. This method includes the following sub-methods: 1. Receive early rewards and education, 2. Adopt a pensive approach, 3. Find a significance for time, 4. Find the significance within the significance.
- Artistic time travel. This method includes the following sub-methods:

1. Obsess over significant location, 2. Identify people with location, 3. Find the people, 4. Identify art with location.
- Time travel by memory and location. This method includes the following sub-methods: 1. Identify a location with more than one time, 2. Travel to the current time, 3. Identify the location with travel, 4. Travel to the other time.
- Time travel by memory and thought, first method. This method includes the following sub-methods: 1. Think about the location, 2. Think about the location changing, 3. Think about change, 4. Think about changing location.
- Time travel by memory and thought, second method. This method includes the following sub-methods: 1. Pick a random thought, 2. Identify the thought with the location, 3. Morph the thought, 4. Change location.
- Traveling by determinism. This method includes the following sub-methods: 1. Establish the knowledge of the past, 2. Establish a causal principle that would create time travel, 3. Cause the past, 4. Live the future.

- Traveling by volition, type 1. This method includes the following sub-methods: 1. Prove complexity, 2. Prove contingency, 3. Prove preference, 4. Prove volition.
- Traveling by volition, type 2. This method includes the following sub-methods: 1. Adopt a dimensional view of space, 2. Imagine that effects occur any where within that space, 3. Reason that determined space is not reasonable, 4. Establish a temporal machine.
- Divine time travel by significance. This method includes the following sub-methods: 1. To have proven significance, 2. To have dynamic significance, 3. To be granted dynamic exception, 4. To exercise significance.
- Divine time travel by mandate. This method includes the following sub-methods: 1. To have a proven exception, 2. To be granted exception, 3. To be granted exception by law, 4. To exercise the law.
- Time travel using magic, method 1. This method includes the following sub-methods: 1. To argue about mortality, 2. To argue that one is made of time, 3. To argue that one

dies by traveling, 4. To argue about travel.
- Time travel using magic, method 2. This method includes the following sub-methods: 1. To argue that magic is old, 2. To lose magic, 3. To gain time, 4. To gain magic.
- Time travel backwards using magical enchantment. This method includes the following sub-methods: 1. Imbue an object with age, 2. Get older, 3. Un-imbue the object, 4. Time-travel backwards.
- Long years using magical enchantment. This method includes the following sub-methods: 1. Imbue an object with youth, 2. Prevent the object from aging, 3. Later, imbue the object with wisdom, 4. Gain the youth of the object.
- Time travel using infinite contingency. This method includes the following sub-methods: 1. Establish an axis of contingency, 2. Assess: all experience is contingent, but not absolutely contingent, 3. Abbreviate existence as non-contingency, 4. Act so as to be contingent.

TELEPORTATION

Results= Fin, Eff= Fin, Diff= Results - Eff no abs notat verified: (Core).

[If efficiency is less than the result, the difference is positive. If efficiency is greater than the result, the difference is negative... Teleportation: Large difference or large efficiency or small distance. This can mean high energy at long distance or short distance.].

TELEKINESIS

There are no scientific breakthroughs with telekinesis, just some people that are slightly better than average at balancing objects, or who might even sometimes cause them to move with their minds.

In the rarest cases scientists assume its just luck. So much luck that its impossible to prove one way or the other from a scientist's point of view. They defer to randomness.

PHOTOS OF OSTENSIBLE TELEKINETIC EVENTS

TELEKINESIS PHOTO 2

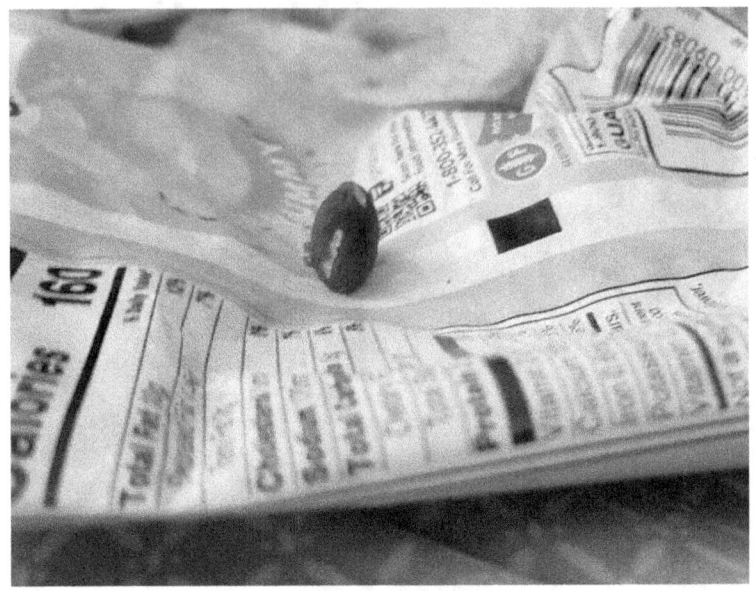

(…)

NATHAN COPPEDGE__IS A PHILOSOPHER, ARTIST, INVENTOR, AND POET WHO LIVES IN NEW HAVEN, CONNECTICUT.

SOURCES: Virtually *all content in this book was sourced from Nathan Coppedge's work.*

www.ingramcontent.com/pod-product-compliance
Lightning Source LLC
Chambersburg PA
CBHW071422210526
45465CB00001B/490